THE LAST ECLIPSE

Planetary Change
and
Universal Being

THE LAST ECLIPSE

Copyright 1996 by Medicine Bear Publishing

First Printing   August, 1996

Medicine Bear Publishing Co.
216 Paseo dell Pueblo Norte
Taos, NM 87571

ISBN 0-9651546-3-7
Library of Congress CCN- 96-094618

Printed in the USA

Cover Design and Graphics by Pat Hayman
Cover Art by the Purcell Team

This book is dedicated to all the children of the universe who live on the blue star of Earth.

Introduction

'The Last Eclipse' is an honest attempt to address some of the mystical and prophetic happenings around this planet and the solar system in recent years...The crop circles, increasing UFO sightings and the discovery of pyramids and stone faces on the surface of Mars suggest that we are awakening to other dimensions of reality. And evidence is mounting to support the idea that human beings have both past and future connections to other planets in this solar system. Humanity is not alone and is being guided rapidly towards universal understanding.

The collective human soul has come from an amazing past and is heading towards an even more amazing future. Our experience on Earth is only but a moment in the cycles of existence and in the destiny of the human seed. We are living in a transitional age that has been foretold by hundreds of mystics and prophets down through the ages. There are indications all around this planet that cataclysmic changes are upon us. We have no alternative now as spiritual beings other than to raise our awareness about who and what we really are and where we might be headed.

There is no death anywhere in this universe, only change through the process of destruction and re-creation...all different forms of the ONE process. Human existence is a continuous process of cycles and spirals moving towards or away from our Creator. This universal aspect of being is embodied within our material 'selves' every time we breathe. We choose and create our paths in this world. There is no cold and punitive 'God' sitting on a throne waiting to send us into fire and brimstone. Human beings create their futures after this life by how they live right now in the moment. The seers, shamans, mystics and medicine people from all cultures have been telling us these things since the beginning of time. It was the repression of the feminine and intuitive aspects of being that separated modern humans from their Creators. The Goddess has returned, though, to claim her place with the God.

Thus, I cast this book out on to the sea of knowledge and discovery and hope that it may bring a little spiritual integration for even a few......

Chapter 1

Eclipses
Omens of Change

One of the most elusive and mysterious subjects in astronomy and astrology today is the subject of eclipses. Only in recent years have modern astronomers and astrologers begun to understand eclipses and their many interrelated cycles. There is evidence all over this planet that some ancient cultures went to great lengths to record and understand eclipses as they believed that they were related to global change and to the entrance on this planet of higher realities. Humanity recently experienced a great and very unusual series of eclipses that have, in my opinion, foretold all the global change that we are now witnessing on Earth and in human consciousness. To understand the importance of the recent eclipse events, we must first understand a little about eclipses in general...

There are always at least two solar eclipses a year on this planet and, in some years, there can be four or even five. By far though, most all total, partial and annular eclipses never last more then three or four minutes. A **major** total eclipse can last seven minutes or a little more but these are very rare. Three major eclipses have occurred in this past century: 1955, 1973 and 1991... all in the same constellation. An eclipse event such as this has not happened for thousands of years and was

foretold by the Mayan Calendar. The importance of these three major eclipses cannot be
stressed enough. They were of maximum
magnitude and all happened in less than one
generation. Astronomical calculations have
shown that it has been at least three thousand
years since such rare seven-minute eclipses
have occurred so close together and that it
will be thousands of years before it happens
again. The Mayans were one of the very few
cultures to have even come close to accurately
recording and understanding eclipse cycles.
According to the Mayan Calendars, these
eclipses signified the end of an age - the end
of many cycles all at once. Maya in Hindu is a
name given to the material world. Mayan
prophecy foretells the end of this material
dimension.

The interesting thing about these three rare
eclipses is that they all took place in the constellation of Gemini. The Gemini God is Mercury; the winged messenger. According to the
ancient Greeks and Romans, Mercury ruled
commerce, transportation, thievery, cunning,
communication and messages. In this century,
we have seen an explosion in all of the above
and especially in all forms of communication
and travel technology. Human beings have
developed winged travel and incorporated

the advanced communication technology of computers into radios and dashboards of all travel machines, automobiles to airplanes. Gemini is the sign that rules the mind. Its element is air and its energy is multiple. Closely related to the muse and the nine daughters of Zeus who ruled over the arts, sciences and music in Greek mythology, Gemini is the entertainer. Just about the time of the first eclipse in 1955 is when a major shift occurred in many forms of mass communication (television, for example) relating to the visual arts, music and other forms of entertainment. Today, humanity is in the midst of a virtual explosion in the entertainment and the visual art industry.

Gemini is also associated with our hands, arms and lungs. Gemini relates to the atmosphere and the air that we breathe. This whole era of cigarette smoking, an activity of the hands and the lungs, is a product of Gemini transformation. On a larger scale, humanity has drastically altered the atmosphere and the quality of the air that we breathe forcing us to become more conscious then ever before about the importance of a non-toxic oxygen supply. Atmosphere levels of pure, clean oxygen are decreasing every year which is affecting the immune systems of all living creatures. Without healthy levels of oxygen, our blood cannot regenerate and stay healthy. All of the aspects of life attributed to the sign Gemini have been "eclipsed" by great change over the past 40 years and are beginning to change even faster.

Look at the computer industry, for example. It is changing so fast that the controlling political forces in the world cannot stop it. A real communication revolution is about to unfold on this planet thanks to the Internet and the WWweb. It was said in ancient times that the winged God of Mercury could travel through the air and communicate telepathically. Could it be that this ability to communicate with the mind *only* is possible for human beings on a massive scale in the near future? Could it be that the return of this these particular eclipse cycles have brought back the "winged messenger"? The interesting thing about the Internet is that it has speeded up our ability to communicate with others around the world...faster than even the telephone lines themselves, once you are linked on. What is next? Computers are an extension of the human mind and we could be entering a realm of instant manifestation of our thoughts into reality...an exciting but dangerous place. The negative human traits like dishonesty, thievery and cunning have also been activated by these eclipses and humanity is struggling with the most massive scale of deceptive, criminal behavior ever to manifest on the planet. The energy and impulses of change released by the eclipses were intense but neutral. It has been available to all human channels and can be used either way: for our higher good or for our demise. Both extremes are evident in the world today.

Hold on to your seats, though, for there is a deeper twist to this story: During the July 11th, 1991 eclipse which, at its peak, occurred directly over Mexico City and most of the Mayan temple sites, a very mysterious and incredible event took place. Middle class Mexicans, much like middle class Americans, had become obsessed with home video cameras. More then 500 independent cameras that were fitted with special lenses to view the eclipse recorded something besides the moon moving across the disk of the sun. They recorded a fleet of UFOs hovering over the city. There were 17 independent video films of the same UFO from different angles! And ever since that day, there have been thousands and thousands of sightings, many of which have been documented on film. The sightings are still increasing with UFOs being filmed off the coast of the Yucatan going in and out of the ocean. What are they doing going in and out of the ocean? The 91 eclipse over Mexico City was the largest, single mass sighting of UFO phenomena ever to be legitimately documented or to be reported on the front page of so many newspapers in one country. Of course, we Americans did not hear much about this event or about the continued sightings and encounters now taking place in Mexico. What is our government really trying to hide, anyway? If you go to Mexico City today, people have their eyes turned towards the heavens every chance they get, and there are other strange and mysterious things beginning to happen in that country which seem to be related to the presence of an alien force. One has to ask if there is some preparation by higher forces for a mass shift in consciousness going on down there?

If there is a UFO connection in times of calamity, then it would make sense that the Mexican people are being closely watched. After all, Mexico City is one of the most populated areas on Earth and the Mexican economy is on the verge of complete chaos and collapse. Not to mention that American businesses and corporations have been setting up factories in Mexico for years that are dumping toxic pollutants everywhere around the Mexican countryside with no regard for human, plant or animal life. They have gotten away with it because the almighty dollar has such a strangle hold on the Mexican economy. Have you ever been to Juarez...across the Rio Grande from El Paso, Texas? I have and I'll tell you it's a tragic sight to see little children covered with open sores, choking and unable to breathe normally (yes, choking breath, the negative Gemini thing) trying to play in the back yards of smelters and chemical plants that have been banned in this country...many of them whose owners have criminal charges against them here in the States for their past environmental destruction. And, in southern Mexico, Wall Street insiders and Fortune 500ers have pushed the Mexican Government to further repress the Mayan-Agrarian revolution in order to help stabilize and protect their investments. Considering that the last powerful eclipse passed over the sacred Mayan sites, it would be wise for the governments of the world to have second thoughts about continuing to mount military campaigns against a peaceful, Earth connected people. If there is going to be extraterrestrial intervention on this planet, I would guess that Mexico would be one of the first places. Mexico City could experience an 8.5 magnitude earthquake at any time. A million people could lose their lives in an instant. What would happen to all those souls?

I can remember vividly what happened to me at the age of fifteen when I was pulled out to sea while body-surfing on the coast of Southern California after getting caught in a rip tide. As my lungs filled with water, the panic and struggle faded into soft waves of peace. The essence of my being - my spirit or soul, ascended right out of my body into another dimension that seemed to be there all the time. I was eventually revived on the beach and I'll never forget the feeling I had when my soul re-entered my body. The experience of freedom and universal oneness collided with mass and gravity. If a million souls were to leave this planet in one instant, I am convinced that there would be a great wave of spirit away from matter. There would be a great rush of souls into the next dimension and there would need to be some kind of stewardship and guidance for these souls.

It appears that there are forces, very advanced beings, now surrounding our planet.

Just because we do not fully see them with our rational minds does not mean that they aren't here. There are many unexplainable things happening around this world, one of these being the increasingly more complex crop circles and designs being left in fields all over the Earth. Research into this strange phenomena is providing some very interesting facts. First, there is absolutely no rational explanation for these crop imprints. Governments and news media at first tried to discredit them with the help of greedy hoax perpetrators but they were quickly silenced when the designs became larger and more complex each season. More and more ancient and mathematical correlation begin to appear in areas larger then 3 football fields over night and in many different countries at the same time. Despite what we are being told by our governments, these crop circles are intensifying and diminishing our grain harvests. There are many who believe that these symbols are messages for human beings. Why is it we tend to think that we are the center of the universe? It could be that the forces surrounding this planet have come up with a very ingenious way of communication with each other, leaving messages for incoming craft or recording the timing of future events on Earth. Perhaps it does not matter if we understand these symbols at all. What does matter is that we pay attention to the fact that there are higher forces at work on and around this planet and this should have some bearing on how we live our lives.

Gemini rules the mind and our mental capacity to store and recall massive amounts of information - something that we have in common with quartz crystals. All of the modern communication technology has been made possible through the discovery and use of quartz crystals and their artificially grown counterparts - silicon chips. Discovery is another keyword of Gemini and things really changed in the scientific world when researchers discovered how to grow artificial quartz crystals and mass produce silicon chips. They are now in virtually everything that we use. We are living in a crystal technology. The energy of the Gemini eclipses manifested in the understanding of the power of crystals, and humanity has applied this power to the material world. What possibilities may lie ahead for humanity if we ever applied crystal technology to spiritual development? There is a connection between the use of crystals and the winged God Mercury in ancient mythology. Perhaps we are just re-awakening to ancient knowledge.

Eclipse cycles have provided some very interesting correlation with human development on this planet. Usually any one specific area of the Earth's surface experiences a total eclipse of the sun directly overhead within a 450 year period. However, one area could experience two total eclipses within a short period of time or no total eclipses for two thousand years because the eclipse cycles are very complicated. Eclipses are actually evolving according to the gravitational stability and planetary movements in the solar system and the galaxy as a whole. Specific cultural areas that have experienced frequent eclipses directly overhead have gone through sudden and revolutionary change. The connections between human experience on Earth and eclipse cycles is very revealing. Every single historical period of revolutionary and cultural change can be correlated with a major solar or lunar eclipse! There are very large planetary and galaxy cycles that the Mayans were able to record with their exact calendars. Eclipse cycles are related to the orbit of the moon around the Earth and they are indicators of the planet's axis and rotational stability. The Mayans took the natural 13 month lunar calendar to be the true planetary standard as the moon makes 13 full orbits around the planet each solar year. They were able to understand much larger, universal cycles by recording the moon's movements across the ecliptic points. There is always a direct line from the equator of the Earth to the sun. Every once and a while the moon's orbit crosses this line partially or totally eclipsing the sun. The Mayans understood that eclipse cycles changed as the Earth's axis shifted when it was influenced by other gravitational and galactic forces. Thus, they were able to predict the cycles of change on this planet. Even though the Mayans developed a solar calendar, it was the *lunar* calendar that foretold planetary and spiritual change. The Western world has been stuck in spiritual and astronomical ignorance ever since the introduction of the currently used 12 month Gregorian-solar calendar which was based only on religious and political needs of banking and collecting taxes. Without the understanding of the natural 13 month yearly lunar cycle which is the true universal standard, how could modern societies begin to understand eclipse cycles or the planetary changes they foretell?

The moon revolves around the Earth and the planets around the sun. But the sun also revolves around the center of this galaxy. The whole solar system is moving at the speed of light all the time through space, but we are not conscious of this movement. The Mayans and their Toltec ancestors, with centuries of careful observations of eclipses, were able to detect changes in energy and gravity as the solar system passed through certain areas of the galaxy. This would often affect the Earth's axis alignment and the rotation of both the Earth and the moon which, in turn, would effect eclipses. They found that some areas of the galaxy have different gravitational frequencies which causes the gravitational balance of our solar system to change. They also found that some areas of the galaxy have more comet and asteroid debris. These areas were dangerous for the Earth and other planets. The Mayans actually recorded a major change in the alignments and orbits of planets around the sun in ancient times! There is also some indication from the Mayan calendar that we are now entering an area of the galaxy that has a higher energy vibration more directly linked to the Great Creator force at the center of our universe. This area is intense and, from what I can decipher, has many more races of beings flying around because this energy field is a preferred realm.

Back to eclipses...it has long been believed by many cultures, especially the astronomically advanced ones like the Mayans, that a total eclipse of the sun would activate the deep unconscious and sacred energies of both nature and human beings. New levels of consciousness and experience could be energized depending on the needs of the planet and its living beings in the specific area in which the eclipse takes place. The ancient astronomers and mystics carefully plotted eclipse cycles because they realized how they foretold change in the deeper layers of existence. The three powerful eclipses of this century have brought the most rapid change in consciousness that human beings have ever known. The 1991 eclipse was directly over the most sacred Mayan temple sites in Mexico and the central theme of the Mayan prophecies is the transformation of human consciousness on a level never before experienced on the planet.

UFOs, real or not, represent advanced travel technology and communication which is governed by Gemini. UFOs are real on some level of existence, whether we experience them in the material dimension or the non-material. They could represent our own cosmic future just as a radio or television set was first a vision or dream in someone's mind.

The most basic elemental nature of Gemini is change. Humanity is faced with change beyond our wildest dreams and plans. I heard a good joke the other day; "If you want to make our Creator laugh, tell him or her your plans"...It amazes me how we 'modern-civilized' people make so many schemes and plans for the future when we are not even in tune with the thirteen moons that make up the true Earth-universal year. In 1991, I was very fortunate to have had some time to fast before the July eclipse and to have spent some time with Hopi Traditionals. Their prophecies of "Earth Purification" and coming Earth changes are very close to and actually are an extension of the Mayan prophecies. Many things that they have revealed to interested people have come to pass already. I believe that the next five years on this planet will tell it all.

The three major Gemini eclipses of this century are the climax of a major cycle which will affect all of humanity. But every person is born within a smaller, more personal eclipse cycle. These cycles are the partial and annular eclipses that are less intense and softer. The

Creators designed the cycles of planetary change in such a way that conscious life could experience change and evolution in gradual steps. But every once in a while, our evolution or development is speeded up with a series of powerful eclipses. This is exactly what happened recently. I believe that the area of the galaxy that we are entering is so intense that human beings needed to be prepared suddenly for the changes in consciousness that will result from the exposure to higher energy vibrations and different gravitational frequencies. Since these recent 7 minute eclipses were in the constellation of Gemini; the realm of the mind, it is likely that the collective consciousness of humanity is being prepared for a grand re-entrance of spirit. We are being prepared to meet our Creators. Every individual person has a place in this drama of change...has a role to act out and fulfill. Understanding the meaning of the eclipse cycle in which we were born can increase our awareness of soul purpose and speed up our spiritual development.

No individual astrological interpretation is complete without a study of the personal eclipse cycle. The last solar and lunar eclipse to occur before your birth determines your eclipse cycle. To understand the nature and sign of your cycle is to understand the process of change and transformation in your life. To do this you must consult an experienced astrologer who has knowledge about eclipse cycles. As an example of how this works, let me briefly relate some information from a recent chart that I was able to interpret. Richard, a 45 year old friend, was having a lot of problems creating a stable relationship in his life. He had been married 3 times, the longest marriage lasting 1 year. We cast his chart together and focused immediately on the seventh house of relationships. Behold, he was born with the sun in the seventh house two days after a strong partial (4 min.) eclipse in Scorpio. His eclipse energy was very intense and direct. His soul came into this life destined to experience dramatic changes in relationships until he changed the way he projected himself. Eclipses always take place on or near the moon's nodes, the ecliptic points of the moon's orbit. These are the two places, north and south of the equator, where the moon crosses in front of the sun. This north-south nodal axis in astrology represents the line of destiny in a birth chart - the doorways of soul entrance and exit in each life time. We enter through the south node and exit through the north. The south node represents all that we have come from and what we already know. The north node represents what we must learn and change.

Richard's eclipse took place at the south node in his seventh house. He never had a problem manifesting relationships. He could manifest hundreds of them. The problem was that he could not find any peace in relationships or manage to develop one of them into a happy and loving experience. Well, the line of destiny pointed to his first house where the north node was located. His karma or destiny in this life was to change the way he projected his personality on to others. The north node was in Taurus and was not well aspected by Mars and Pluto. This was a clear indication that he had the tendency to be selfish and stubborn (a negative Taurus trait) and was overly possessive with money and material things. Until he realized that it was he creating the disruptions in relationships and was willing to change this aspect of his personality, he would never have a stable relationship. And he would not exit this emotional and material plane of being that his soul was experiencing until he changed. Once he became fully conscious of this behavior, he quickly entered counseling and made great progress. Most of the time human beings act unconsciously from past-life habit patterns until they somehow are made aware of their patterns. The pattern of conflict and change in Richard's life became obvious when we looked at the cycles of return eclipses in his chart. On his 27th birthday, there was an eclipse in Scorpio at the exact degree of the one near the time of his birth. Richard confirmed that it had been a time of great inner conflict and change and that he had lost the one true-love of his life because of his selfish and stubborn behavior. Had he been aware of his past-life habit patterns before his 27th birthday, things may have turned out different. Astrology is simply a tool that we can use to become aware of our soul-purpose in this life, and eclipses are the omens of necessary change.

Astronomy and astrology have taught me that there is a natural progression to higher existence in the galaxy and that every aspect of this universe is interrelated and part of a whole process. Don't ever fear change. There is a beautiful future ahead of us. We are creating every aspect of our being and experience in the material world with our minds. We must be careful of what we put in our minds and of what we think, especially in this new area of the galaxy because we will tend to manifest what we have in our tissues of our thoughts. That is the message of Gemini. We can change, create or destroy the world with the power of our minds. What a tremendous responsibility befalls us!

The message of eclipses is change - change on the personal, individual level and on the collective, planetary level. The Olmec, Toltec and Mayan cultures spent thousands of years studying and recording eclipse cycles. From these observations, they were able to understand more about the nature of our galaxy and the larger cycles of the sun's rotation around it's creator...a much larger sun! All of the astronomical knowledge of 'modern' consciousness falls short of the knowledge possessed by the ancients. Eclipses are omens of change. There may not be another series of seven-minute eclipses for hundreds and thousands of years. I believe that we must pay attention to these last major eclipses and we must pay attention to what happens in Mexico...for whatever happens in the consciousness of the Mayan people and in the Mayan highland societies will likely happen to the rest of humanity.

Chapter 2

The Prophecies of Edgar Cayce

"Oil and Crustal Displacement"

There lived, not long ago, a great "sleeping" prophet who devoted his life to others in order to teach, heal, redirect and awaken their souls. He revealed to many people how to cure their often "incurable" illnesses and how to nurture spiritual awakening. He taught many people how to transcend their struggles in this life by understanding that they had lived before and would not be here on this plane of existence if they had finished their soul journeys and lessons here on Earth. He often could see into the future and warned of the irreversible harm being done to the planet by modern society. This sleeping prophet is gone now. He lived humbly and he left only the material connections to life which had been necessary to sustain a simple, peaceful existence. All of the knowledge and information left by this man live on though, recorded by hundreds of people as it came through his psychic trances and a powerful openness to the mystical and unconscious realms of being where time and space have no reality. This man was Edgar Cayce. He is considered by many to have been the greatest psychic and medical diagnostic seer of the century.

Just what does Edgar Cayce have to do with the subject of 'Oil and Crustal Displacement'?

The truth is that Cayce would often see glimpses into the future of humanity during his trances. He repeatedly saw the Earth going through tremendous geophysical up-heavals at the end of this century. He warned of sudden catastrophes in which many people would be swept into the afterlife...waves of mass human destruction that would be ac-companied by rearranged continents. Places that are now cold and ice-bound would be-come more tropical and viceversa. Many times, Cayce would repeat the following words before coming out of those particular trances: " In the late nineties and into the next century, there will be large scale crustal dis-placements on Earth caused directly by man-kind". Cayce was never conscious of the information he was giving to people wit-nessing his trances. He always had to be informed or listen to the session tapes. Many of the visions that he had during these psychic moments actually went against his personal and conscious ideas about the world and the future of humanity.

Now what is a 'large scale crustal displace-ment' anyway? Well, it appears that the pum-ping of oil in such massive quantities from the crust of the Earth is causing the weight of the planet's surface mass to change dramatically.

Weather patterns are changing radically now as water needs to be distributed around the planet to create stability and balance, often resulting in the deposit of millions of tons of water in places that normally get very little, while in other places, water is withheld. The world's oceans and atmospheric water are the main points of checks and balances for the Earth's surface gravitational stability. By far, most of the oil that we use has been and is being pumped out of the land masses of the northern hemisphere. The droughts in Africa, Australia, and South America are evidence of a shift in atmospheric water. More and more water has to be dumped into the northern hemisphere. The 1993 flood of the Mississippi delta is a perfect example, as well as the recent floods in Europe, Asia and China. Ice and water concentrations at the South Pole are showing signs of decreasing as well, but water can never really replace the oil. Water cannot filter into the underground chambers left by the removal of oil in most cases and, if it does, ground water is displaced from previous stable concentrations. Add to this that we are pumping even more water every day out of the Earth's crust, and you can be sure that the surface balance of our planet is changing rapidly.

This is the real problem: ultimately the oil that we remove from the ground is burned into the atmosphere with very little of it ever returned to the crustal mass of the planet fast enough to offset the imbalances created. There are estimates that we are burning 70 million barrels of crude oil a day around the world. One barrel weighs about 500 pounds or so. That translates into an almost incalculable number. Thus, the surface mass of the Earth is rapidly, decreasing creating a whole different land mass-water relationship. Deep under the Earth's crust, the molten lava flows must also shift to compensate for surface density changes and gravitational balance. What does this mean for humanity in coming years? I believe that the potential outcome is clear.

The Earth is currently balanced very precariously at a 23 1/2 degree tilt on its rotational axis as it revolves around the sun within the great gravitational belt that extends into the solar system outward from the core of the sun. With the combination of the moon, planets, and all the other gravitational effects, the Earth has maintained its delicate balance for thousands of years - ever since that last major gravitational shakeup which resulted in global floods and volcanic upheaval. It is the mass of the Earth in the proportions given by nature in relation to all the solar system's gravitational fields that determines the balance and stability of the axis tilt and the north-south pole alignment. If we change any of the mass ratios of the Earth, it only stands to reason that a major shift in planetary balance will follow. Even a minor shift of the Earth's polar axis would be cataclysmic; disastrous for human material existence but regenerative for Mother Earth. There is very conclusive geological evidence that the Earth has gone through such changes in the past and that the polar axis has shifted dramatically before. Large mammoth beasts have been found under the ice at the South Pole with undigested plant life still in their stomachs. There must have been one very sudden axis shift to have frozen a whole tropical forest and its beasts in a matter of hours! There are many theories about the cause of past shifts: great meteors hitting the Earth, changes in the sun's mass caused by explosions on the surface of the sun, stressful planetary, electromagnetic and gravitational forces, or the possibility that advanced civilizations misused energy technologies. Chinese history recorded something of the latter. Their written history goes back farther than that of any culture on the planet. There are descriptions of mass destruction and nuclear fires in the sky in ancient Chinese records. Whatever be the case, there is no doubt that the Earth can change its axis direction and rotation speed at any time now.

Edgar Cayce never considered himself a prophet. It was his family and friends that recorded his psychic messages. Could it be that Cayce's warnings were in accordance with the prophecies of various cultures such as the Mayan, Native North American and the Christian? I believe that they were. It is interesting to ponder on the Christian psychic revelation left for us by St. John. According to his words in the last chapter of the New Testament, he states that, in the end times, "the dead shall rise from their graves" and "the beasts shall rise from the oceans". In reality, oil is the dead remains of past life: compressed plant, animal and human remains. Sudden catastrophes tend to bury large deposits of these remains in low places in the form of sludge and peat. As this material is buried over time, it is heated and compressed into oil. It takes about 100 feet of compressed vegetation and animal mass to make one level foot of peat. And it takes 100 feet of compressed peat to make one level foot of oil.

There have been oil reserves found that are hundreds of feet deep. That's a lot of compressed, once living matter! That is all the life from the past on this planet - all of the dead. Modern society now feeds off these remains of past life and is, literally, "raising the dead" into the atmosphere through the burning of oil. We have even created giant machines (beasts) to destroy the Earth that are not only fueled by the dinosaurs, but actually resemble them. Take the monstrous uranium cranes on the Navajo reservation in Arizona for example, or any of the armor-plated bulldozers and excavators plowing under the world's forests.

Many Native American cultures have ancient prophecies that tell about a time when Mother Earth is going to regenerate and purify herself from the plagues of human beings. What better way to purify the planet than an axis change? The New Testament speaks more than once about a "new heaven" and a "new Earth" after cataclysmic upheavals. If the Earth's axis where to shift, the sun, moon and stars would rise and set in completely different patterns...in essence, a "new heaven". The land masses of the planet would be arranged totally differently and we would have new north and south poles. Great sheets of ice would form quickly over now inhabited areas while the ice in the current polar regions would melt, creating super currents of cold, clean water in the world's oceans...a "new Earth".

I am not writing this chapter as some type of doomsday report or prophecy of disaster. What I have done and continue to do is to gather information - facts - from the sciences and many of the mystical prophecies of cultures around the world. Along with my own objective and intuitive observations about current Earth changes, I have tried to weave all of this information into a holistic picture. I believe that many of the world's problems are a direct result of political forces and world leaders not wanting to see the larger picture, nor have they placed any value on the tremendous wealth of information passed down to us by psychics and seers.

There is no doubt in my mind that the Earth has already begun to move towards an axis and rotational shift. But what motivates me to write these words is something more important than the subject of planetary change. There seems to be nothing that we can do to avert the dramatic Earth changes likely to unfold in coming years because the use of oil is actually increasing despite all the well researched information on clean and safe alternative energy sources that are available to us. Oil has become a symbol of the problems and sickness within the collective soul of humanity. Forces of greed, materialism and destruction surround the aura of oil. All we can do as individuals is to develop our spiritual values and focus less and less on material gain and comfort.

I became aware of just how bad the forces of greed and destruction really are a couple of years ago while in New Mexico. I had stumbled on to a small group of researchers and scientists who had, 10 years earlier, discovered how to turn water into hydrogen gas. They were able to conclusively prove that it was practical, cheap and environmentally safe for the conversion process was done with solar panels. They presented their findings to the US Congress with estimates of 5 years for the whole conversion process of every motor vehicle in the country if the project was funded. The only bi-product of hydrogen gas is steam...water that would recycle safely in the world's ecosystem. Well, you can guess what happened. These scientists suddenly became silent. Were they targeted by powerful oil lobbyists and government insiders? Were they were threatened into silence or bought off? And now, our oil saturated society is heading for a drastic plunge into chaos. I estimate that we will be entering this plunge in the near future. It would not surprise me if we experienced a first wave of movement in the continental plates in the next five years. And we will have to face the economic chaos that is developing from the twisted minds of oil barons.

It has to be this way, I guess. More than likely, many of the ancient prophecies will come to pass. It seems that humankind must learn the hard way in order to learn at all. If it is our ultimate destiny to develop into Gods and Goddesses or universal creators, then we must have the experience and knowledge of environmental management. We must have experienced the consequences of environmental destruction. Perhaps this is part of some kind of universal growth process. I have always had a hard time with the concept of a judgmental and punitive God. We do have some control over our fate and we do have free will. Humanity has the potential to change and redirect its destiny. Prophecies about the future and myths about the past have always served to warn against certain collective forms of human behavior. On this spiral of evolution, we can never be quite sure

if we are heading up or down. After all, it's all
relative to some aspect of stability and who
knows?...prophecies could be myths from our
future which we could actually be moving
away from. When we are finally aware of
more universal concepts, we may just find
that we were headed towards the "planet of
the apes"! In any case, prophecies have
served to direct all people to live more holistic
and humanitarian lives. We cannot continue
to blame the destruction of this planet on
"God's wrath". Human beings have been entrusted with the garden of Earth in order to
develop towards more responsible and caring
beings with a spiraling spiritual consciousness. If we fail to do this while on this planet,
let's hope that we get another chance somewhere else!

If our focus in this life is on money or material things regardless of the ecosystem, animals or other human beings, then any future
Earth changes will likely be a terrifying experience. Our challenge is not only to change
our consciousness about Earth stewardship,
but to change our ways of being. If we make
such efforts now, the potential cataclysmic
Earth changes will have no real permanent
effect on our souls. We will just continue to
sail into other dimensions with the awareness

which we have acquired during our stay here on Earth. Our future after this life is determined by the spiritual balance of our souls - not by any balance of material wealth or status. It is determined by how we live and act in the present realm of being. We create or destroy our "heaven" right here on Earth. Throw out those old concepts of a heaven with golden streets somewhere up in the sky. We had better clean up our ideas and behavior right here on this planet.

Human beings have created the present state of existence in the world. We are the creators of "heaven" or "hell". We are creating our own futures after this life by how we live in the present. To live a life of careless destruction and degradation of this beautiful Earth garden would surely return us to a hellish existence in the afterlife until the lessons of love, honor and respect for all life were learned. It seems to me that we are still trying to live in the past. Oil is the substance and symbol of all we need to leave behind. Did we really have to dig up and pump up the remains of ancient beasts to fuel our movement into the future?

Chapter 3

Back to Mars

I have to wonder how many people are aware of the raging controversy among the astronomical and scientific communities these days about some photos of the Martian surface. The photos in question are of the Cydonian region on the red planet that were taken by the 1976 USA Martian Observer Probe. Hundreds of thousands of photos were sent back to Earth by this probe and it wasn't until a few years later, when photos of the Cydonian region were taken from storage archives and computer enhanced, that a massive stone humanoid face was detected staring out of the reddish sand: a structure larger then any of the sculpted stone ruins here on Earth. And not far from the face there were pyramids, many of them all seemingly aligned in mathematical precision towards the face. It has since been calculated that if one were positioned in the center of the pyramid city looking towards the Martian east at the time of the equinoxes in ancient times, the sun would have risen directly out of the mouth of this face...a very advanced engineering achievement!

Leading one of the first research teams to study the Cydonian photos was Richard C. Hoagland. Mr. Hoagland is no stranger to the scientific community. In 1965, at the age of nineteen, he became the youngest curator ever at the Museum of Science in Springfield, Mass. Shortly thereafter, he was regarded as one of this country's most respected authorities on many astronomical projects: NBC consultant for Surveyor I and other moon landings, CBS consultant to Walter Cronkite about the Apollo Program and much more, the list being too long to include in this chapter. Now the founder and president of the Mars Mission based in New Jersey, Mr. Hoagland has devoted the last ten to fifteen years of his life to the study of the Martian ruins. In his book, "The Monuments of Mars", he has chronicled the whole story from the discovery of the "surface abnormalities" on Mars to the now recognized advanced mathematical technology of the Cydonian pyramids. He has also chronicled the attempt by NASA and the American Government to hide these discoveries from the general public as well as their attempt to discredit hundreds of reputable scientists, philosophers, physicists and astronomers from all over the world who are researching these important discoveries. In the fall of 1993, when the new Mars Observer was beginning to enter the final orbital phase around the planet, equipped with cameras that had 50 times more clarity and power to view the Martian surface, the mystery deepened. According to NASA, the Observer vanished...or did it?

I must admit that I was very skeptical when first reading about the discoveries on the surface of Mars. But, after 4 years of active research, I have no doubts that there are stone ruins on the red planet indicating past inhabitancy by advanced humanoid beings. I studied the photos from the Cydonian region but I am no expert and I tend towards mysticism in the first place. So I started calling different universities and scientific institutions that were involved with the Mars Project and came upon Stanley McDaniel, Professor Emeritus of Philosophy and Epistemology at Sonoma State University in Arizona. He had just finished a year of research linking a team of independent scientists from around the world, which culminated in a 200 page "McDaniel Report". This report, presented to the U.S. Congress and NASA, documented a 17 year independent scientific investigation of the Martian ruins along with the 17 year deliberate cover-up by NASA.

What did the McDaniel report find? First and foremost, the study found that NASA has never carried out one single objective, scientific inquiry into the "landforms" and that NASA has consistently sent false and misleading statements to the press and to members of Congress, hoping to discredit the vast body of independent and reputable scientists that have been dedicated to the Mars project. Of the many "landforms" in the Cydonian region, the face has been the most thoroughly studied and, according to McDaniel, "the photos have undergone the most exhaustive series of tests for the evaluation of digital images originating from an interplanetary probe available to scientists today." The most advanced techniques of image enhancement, such as photoclinimetry and fractal analysis, have been applied to the photos by investigators who are the acknowledged experts in their fields. In every single test, the data has shown the face to be of artificial origin and masterfully created. Many of these studies are covered in Richard Hoagland's book. McDaniel did not set out specifically to prove the reality of the Martian ruins. He left that task to the expert astronomers, geologists, physicists and others. Being a professor of Epistemology (the study of the nature and origin of human knowledge), he had other motives. His report was a formal protest to the deceptive policies of NASA. It is his view that both NASA and the US Government are morally responsible to provide any verified discovery of past or present extraterrestrial intelligence to all of humanity and that any deviation from this moral responsibility goes against the long standing NASA policy of not withholding any important information regarding interplanetary discoveries. Any such discoveries would be revolutionary for the development of human knowledge!

In the fall of 1992, Richard Hoagland addressed the United Nations in New York. For nearly two hours, Mr. Hoagland kept the UN members spellbound with a riveting slide show of the Martian ruins and their seemingly advanced geometry. He was able to relate this geometry to some of the ruins on Earth. This set the stage for many major news networks to prepare for the fall of 1993 when the new Observer was set to enter the Martian gravitational belt. Many of the networks had made preparations for live, national coverage of the incoming photos although NASA was carefully avoiding any commitment to re-photograph the Cydonian region.

On the morning of Aug. 22, 1993, Mr. Hoagland was on ABC's Good Morning America for a short but very important debate with Dr. Bevan French, the official NASA spokesperson for the Mars controversy. The debate did not go well for Dr. French and, at one point, the ABC host simply asked, "Why, Dr. French, don't you just take the pictures and prove these guys wrong?" Just as the show concluded, a NASA spokesperson from California made a call to Lee Siegal, science editor for the Associated Press. The spokesperson, Robert McMillan, told Siegal that something had gone wrong with the Mars Observer the night before....A fourteen hour gap with press communications!

Within a few weeks after the "loss" of the Observer, Richard Hoagland and others received information from engineers within NASA that the Observer was indeed alive and sending photos back from Mars and that the whole project had been turned into a secret "stealth" Cydonian mission. The U.S. Government and the military wanted a safe window period to view the incoming photos before they ever went on national television. In recent years, serious charges of scientific and ethical wrong-doing have been directed at both NASA and Washington by the worldwide scientific community. Why is it that certain people in power do not want the general public to be aware of potentially higher realities? I believe McDaniel has answered that question well. If the American people are fully informed about these discoveries, they are likely to begin questioning all the current theories of human origin, purpose, and political-religious future. The discoveries could actually *unite* human beings on this planet in a quest for universal truth. This is what the government really fears.

The NASA cover-up of the Martian ruins has become one of the most deceptive policies ever to disgrace the American people. So many leading scientists, journalists and educators from around the world were standing ready for the new photos in the fall of 1993...and then they were suddenly told that there was nothing to see. But NASA was really too late to stop the Mars Project from going forward. Many thousands of hours of research had been done and many findings had already been presented. It has become only a matter of more detailed photos to present to the general public. What does this all mean and what are the scientific implications of such a discovery? Mr. Hoagland and others have outlined three possible theories, all of which would radically alter our conceptions of human reality:

Either humanity was on the planet Mars before it cooled, lost its atmosphere and its ability to support material life as we know it...and was somehow transported to Earth.

Or humanity has been visited by alien beings or Gods-Goddesses that built similar pyramids on both planets.

Or there were other beings in this solar system that evolved independent of human beings who just happened to look like us and built pyramids like us.

Whatever be the case, something very powerful is going to change in the collective human consciousness when these discoveries are fully brought into the light.

Something has changed in my conscious mind. My whole thought processes have changed as I question more than ever the established intellectual, political and religious norms of today. It is very important for all of humanity, for our future and spiritual development, that we demand a full investigation of this matter. The truth should never be denied even if it does begin to dismantle our current belief systems. So many strange things are manifesting in this world today...so many new discoveries about our universal past and our universal future. We must not turn our backs and retreat into the Dark Ages when people were forced to believe in the self-serving doctrines of political institutions. We must acquire a taste for the truth at whatever cost.

If anyone is interested in more information about the Mars Project (now the Enterprise Mission) headed by Richard C. Hoagland, write to The Enterprise Mission, 122 Dodd St., Weehawken, NJ 07087. Various books, videos and the Planetary Horizons Newsletter are available.
Email at www.enterprisemission.com

Chapter 4

Planetary Evolution

There are three planets in this solar system, Mars, Earth and Venus, that have a solid, outer crust and have an atmosphere of varying density. All three of these planets seem to be evolving geophysically along similar paths. The surface of Mars has turned to dust, the atmosphere has become very thin and cold, the polar ice caps are evaporating, the rivers and oceans have dried up and the mountains are crumbling. On the other hand, Venus has a very thick and hot cloud laced atmosphere. Venus is hot, primordial, volcanic and virgin in nature, much like the Earth used to be millions of years ago and Mars millions of years before that. Mountains are forming on Venus and as Venus's atmosphere cools, oceans will appear and the matrix needed for material existence will develop. Earth is the middle planet with warmth, cold, high and low pressure systems, all in a dynamic interplay. For hundreds of thousands of years, the Earth has been cooling off and gradually losing its atmosphere. Mars is currently in a state of high pressure dry air with all of its water and oxygen completely evaporated. Venus is in a virtual state of low pressure, element saturated, dense air. Somewhere between the two extreme conditions is Earth.

Life has always existed in the solar system. As the solar system formed and evolved, life has also formed and evolved through the different spectrums, wavelengths of being eventually manifesting on the material plane. A process of human evolution has been going on in the whole solar system, not just on Earth. Conscious life appears to have manifested on Mars long before it manifested on Earth. Mars is farther from the sun and would cool the fastest of the three planets. It would lose its ability to support material life. The geological features on the red planet reveal that there was a time when rich oceans bathed the shores of plush continents and fertile bays. During that time, the Earth was literally boiling. But out of the superheated lake of protein and elements came the infrastructure of nature needed to support plant, animal and human life.

Astronomers have seen evidence that the sun is shrinking and that the molten cores of planets tend to cool over time. Mars began to cool rapidly and quickly lost its ability to support any form of material life. Just about the time of the mass extinction of dinosaurs on Earth, Mars was drying up and material life was dying. As the Earth's geology reveals, there have been great and sudden shifts in the balance of planetary bodies which can send planets into rapid cooling trends and atmospheric change. Could it be that intelligent life existed on Mars and somehow had to be transported to Earth when the red planet could no longer support them? Advanced studies of the NASA photos of Cydonia reveal very strong similarities between the Martian pyramids and the pyramids here on Earth. And the face, mystically staring out into space, has been shown to have a connection to the sphinx in Egypt. Studies have also revealed that the Mars pyramids are better preserved in the cool, thin Martian atmosphere than the Egyptian and Central American pyramids and that all the pyramids on Earth are much older then scientists previously thought.

What happened to the dinosaurs anyway? They vanished from the face of the Earth in a very short time. A few of them survived, evolving rather quickly either into miniature reptilian land creatures such as lizards and snakes or into sea creatures. For a long time after the planetary shift or catastrophe that brought on the mass extinction of the dinosaurs, there was very little food available and, in time, all kinds of new creatures showed up as the plant kingdom recovered.

Was there some kind of divine - higher - intervention in the evolution of this planet in order to prepare for the entrance of human beings into this ecosystem? Were the dinosaurs destroyed purposely by an extraterrestrial force? Surely the vast array of plant, animal and human life that emerged after the sudden planetary change could not have existed without the rich and fertile garden that was created by the dinosaurs. The peat, topsoil, coal and oil left from the long inhabitancy of the planet by the dinosaurs was critical to support the vast sea of humanity that eventually emerged As we have previously found, modern society, in its current material consciousness, would be at a total standstill without the remains of these great beasts and the tropical forests that they fed upon. We are pumping the remains of past life to fuel our present economy. Geologists and anthropologists have not been able to solve the mystery surrounding the disappearance of the giant reptiles and the relatively sudden appearance on Earth of so many diverse species of mammals. The theories of natural selection and evolution collapse during that time-period, for whatever it was that destroyed such a massive and well-adapted species in a matter of days - possibly even hours - must have surely destroyed most all other species as well. At the deepest layers of ice stratums in the Antarctica, there are remains of great tropical forests and their ancient inhabitants. What ever swept over the dinosaurs also brought on the planet's first wave of ice and cold.

After years of research, I am convinced that humanity or our ancestors did exist for some time on Mars and were somehow transported to Earth from Mars and possibly from other places as well. There is more and more evidence emerging to support such a claim and humanity is increasingly beginning to experience the presence and reality of more advanced beings on Earth during these climatic, dangerous times, perhaps in preparation for mass transport in the advent of any global catastrophe. I believe that there is solid evidence on this planet which supports a theory of mass transport sometime in our distant past. High on the mountain plateaus of the Andes is a region that, mysteriously, resembles the desolate, cold, dry and uninhabited landscape of Mars. There has been a lot of speculation by scientists and researchers about the unusual ruins discovered there; but most of the scientific community has, unfortunately, dismissed further study of these ruins and put them into the archives of "unknown". Yet this does not change the fact that they exist. There are signs, guideposts, markers, apparent runways and cultural symbols spanning miles on these high plateaus. From the ground, they are not distinguishable at all. From the air, they reveal an amazing realm of ancient activity. Marks and arrows pointing to different places on the planet are everywhere. Could it be that all of these lines, animal and human figures created on these high plateaus were directional indicators leading to various cultures and climates on Earth? If human beings were transported from Mars to Earth, it would have made perfect sense to enter this biosphere at a place of familiar climate to allow biological adjustment.

Computer enhanced photos of the great humanoid stone face on Mars shows a sphinx-like nature much like the sphinx in Egypt. The pyramidal ruins, as well as the individual designs, are positioned in exact mathematical angles. And there is a massive five-sided pyramid, larger than any on Earth (called the D&M pyramid), that appears to house incredibly advanced mathematical formulas connected to the gravitational balance of the planets and the solar system as a whole. Who knows what we would find inside these pyramids if we could ever get a manned mission to Cydonia. We would likely uncover startling artifacts.

Now the most puzzling question of all is this: If all or some part of humanity existed at one time on Mars, how did we get to Earth? What force brought us here? My research, spanning over twenty years, has convinced me that humanity has always been observed by higher-creator forces and that there was some form of mass transport of human (and possibly animal) beings from Mars to Earth. Archeological, mythological and cultural evidence of divine intervention on this planet by outside forces can be found at all corners of the globe. From the Chinese to small nomadic tribes in Australia and Africa, researchers have recorded myths and found stone inscriptions describing flying crafts, space travel and Gods-Goddesses coming from the heavens. Myths are aspects of oral history transmitted over long periods of time through generations of culture. In my opinion, early myths and stories of human experience should be considered more accurate than later historical accounts of human experience recorded by politically and religiously motivated cultures. The Greeks, Egyptians, Baby-

Ionians, Africans, Chinese, Mayans all had many myths of the Gods and Goddesses with their supernatural power to travel to other worlds.

If there were flying crafts in ancient skies, how does it happen that we have never found the remains of one in all of the archeological research that has been done around ancient stone cities? It is possible that the ancients had a connection with beings that had knowledge of what science now calls the realm of anti-matter. These beings could have visited Earth and directed many aspects of cultural birth around the globe. Travel in this realm of anti-matter would leave no trace in the material world other than in the human conscious experience. This could also explain why we have absolutely no trace of evidence any-where around the planet of how the ancients sculpted and transported the great stones in their structures...sometimes for hundreds of miles and up to 12,000 feet in altitude. We have never found wheels, cables, rafts, tools or leverage objects associated with any of these thousands of massive stone structures even when bones, pottery shards and other cultural artifacts are discovered from the earliest periods of habitation around these sites. There are hints of the truth about humanity's past everywhere in archeological research; but many scientists, researchers and political leaders still fear the idea that the ancients may have had a more direct link with and more knowledge about universal realities and the true nature of universal being.

Before moving to the next chapter, it may be interesting to note that the Earth is about fifty-fifty high pressure and low pressure atmospheric balance at any given time. And the cooling off process on this planet has been dramatic in the past 100,000 years. Humanity is speeding up the cooling off process faster than modern science can understand. There have been many reports about global warming but most people are not aware that artificial global warming would set off a massive ice age ultimately. Artificial warming would tend to produce more clouds and storms. If we combine this with the sulfur, carbon and other elements that "technology" is flooding the atmosphere with, sooner or later the sun's warming rays will be reflected back out to space resulting in a sudden cool-down of the Earth's surface. Similar, natural cooling trends in the planet's past have been detected by geologists. These have usually been set off by periods of intense volcanic activity resulting in ash and cloud-filled skies. The sun's rays get reflected and great sheets of ice advance out from the poles. One thing is for sure, the great areas of lush, tropical life on Earth are gone. They disappeared long ago. And the sun has been shrinking, leading to faster cooling of planetary cores. Mars would naturally cool first, Earth second and Venus last. Life as we know it would naturally have to follow a path of progression towards the sun!

Chapter 5

Forward to Venus

Understanding the mystery of the incredible ruins on Mars is an important task for humanity in coming years, but we must also focus some of our attention towards Venus. Mars represents our past and Venus represents our future. Venus may very well be our next home on the path of material and spirtual evolution. Of course, Venus is extremely hot right now in terms of our physical senses, but a sudden shift in the electro-magnetic and gravitational balance of the solar system could render Venus cooler and habitable in a very short period of time. Beings with knowledge of compatible plant and animal species could quickly turn a virgin world into a magical garden. There are billions of planets out there in the universe and it is likely that hundreds of thousands of them are capable of supporting material life. But we must realize that there are other wavelengths of existence besides the physical-material and that there are forms of life likely to exist on other planets that human beings would currently view as un-inhabitable. On the other hand, the planet Earth is still very much inhabitable. Even after great planetary change, which is growing more likely as each year passes, humanity may well have a few more cycles of existence here on Earth. We may have to pass through some difficult and challenging times while Mother Earth cleanses and purifies; but this planet is far from drying up!

"Modern" science tends to project its material consciousness towards different worlds, expecting to find or not to find realms of existence that fit its conceptual framework. If human beings were to develop their spiritual consciousness to higher levels, they may begin to experience other realms of being that are presently around them. In these realms, physical and material experience would be different and who knows what Venus is like in the spiritual sense. I know that when I was out of my body after drowning in a riptide off the coast of California, I experienced another realm of being which, over the years, has forced me to expand my consciousness beyond its material and physical bounderies. I have learned that human beings will not find in this world or other worlds anything that they are not able to sense in the first place. Humanity has fallen severely short of being able to sense all the realities that are here in this world right now; yet we still expect to find other realities out in space. The truth is that we look right past them, unable to sense their wavelengths of being. This is the message of the mystics. We can develop the abilities to sense

all of the vast expanses of the universe and
wavelengths of existence right here on Earth
and within ourselves. We cannot develop
these abilities if we get caught and stuck in
material consciousness. That is exactly what
the collective consciousness of humanity has
done!

Humanity has intuitively known that there
have been other realms of existence in the
universe. Mythology, religion and philosophy
have attempted to understand these realms
since the beginning of time. It was this desire
to understand in the first place that led us to
get a probe to Mars and photograph its surface. Why should we be shocked at what we
have found? Did anyone really believe that
ancient human beings built those advanced
pyramids on Earth without some sort of
divine or extraterrestrial help? And if we are
just discovering or awakening to the advanced mathematical technology in these
ancient monuments, what does that say of the
past six to ten thousand years of human experience? What does that reveal about the nature of our human ancestors? They certainly
were not primitive cave dwellers as some
researchers would have us believe.

It would be wise to turn our attention to
ancient astrology and mythology. It is here

that we can find more revealing clues to our past and our future. In all traditional astrology, which has been shown to be the oldest science on this planet, Mars has always been associated with the physical being, our material drives and ambitions, war and basic survival. Isn't this exactly where humanity has come from and what our history reveals? Ancient Greeks worshipped Mars (ARES) as the male god of war, anger, shear strength and courage (sometimes without wisdom), wild passions and sexual regeneration. When material life manifested in the form of human beings, the first and foremost task was survival - eat or be eaten. There could be a whole millennium of human history locked up in the Martian pyramids that may contain information about humanity's past. And who knows what might have happened on the surface of the red planet. Perhaps human beings destroyed that biosphere like they are doing to this one. In ancient mythology from many cultures, there were revered war Gods that had superhuman strength and not much regard for nature, animals and human life. These Gods ravaged the planet in search of sacrifices and sexual gratification. Where did the conception of these Gods come from if not from our own past and experience? As humanity has moved forward in spiritual development and towards universal consciousness, we have been learning to tame the extreme aspects of Martian energies, although they have always been present in our collective unconscious and explode once in a while in all their destructive and murderous forms.

Venus, in ancient astrology and mythology, has always been associated with the Goddess of love, harmony, beauty, and spiritual being. Venus represents the ideal of what humanity could create on this Earth and where humanity can go in the future. Venus is feminine and intuitive. Venus is love and trust, forgiveness and cultural union. These are all things that our Martian heritage has feared. Why trust nature, the universe or others to provide for your needs when you can exert your Mars-masculine domination over everything and take what you want? This is the dilemma humanity has been faced with ever since its arrival on this planet. Every time the Goddesses of Venus have tried to create havens of peace and universal love on Earth, the red head of Mars, full of fear and survival ambition, has re-manifested. Nowhere in human history is this classic struggle more evident than in Greek mythology and in the poetic tragedies, a series of plays in which we witness the subjugation of the Venusian realms of being. To the Greeks, Venus was APHRO-DITE-DEMETER-GAIA and many lesser Goddesses. They were powerful feminine archetypes that developed in all cultures around the planet in various forms. Gaia was the mystical prophetess that revealed to humanity the message of universal love, cultural marriage instead of war, healing instead of inflicting pain. She showed the way to the future. Aphrodite was the goddess of love in the material world, goddess of joy, sensuality, partnerships and charity. She also was carrying a torch for the future. Demeter was the Earth Goddess protecting women, young girls and the fertility of the Earth - the harvest of corn, grain, fruit and vegetables. She frowned on killing animals or eating meat as these acts were only an extension from our Martian past that caused men to rape and pillage nature and the societies of women that she supported. Only after experiencing sacred rituals of honor and love for the feminine and nature, were men allowed to enter her secret societies. Demeter was one of the most revered Goddesses in ancient Greece by both men and women.

The Greek Gods and Goddesses have provided excellent examples of the male-female -

Mars-Venus - struggle here on Earth, the middle planet. But there are the same examples in most every culture, the same Gods and Goddesses, only with different names, representing our past and our future. Earth has been the meeting ground for Mars and Venus, two opposite ways of being. Earth is the middle planet and it makes sense that it is the meeting place.

And look at humanity today. The central crisis facing human beings is about finding the balance between the male and the female. The most pressing need of all of us is to honor the feminine-Venus within ourselves and each other. We cannot continue to dominate and destroy the world any longer or we will end up with a dead planet and a corrupt, empty soul.

Venus could be our "new Earth" someday but let us hope that we learn the lessons of love, honor and dignity here on this planet first. I believe that we will and that humanity is finally realizing that whatever we do to our planet's body, we do to our own. Whatever we do to our brother's or sister's heart, we do to our own heart. Earth is the meeting ground of the male and female energies, a world of duality and individualization. The potential product of this meeting is spiritual consciousness and existence on other levels of universal creation. Long ago our Creator evolved into or manifested into the material-physical planes and into the form of human beings. But the Gods and Goddesses that are within us are now evolving back to spirit...back to the Creator. We can detect this process by looking more holistically at our evolving solar system. The highly advanced and ancient stone monuments that are still standing on this planet and Mars after thousands and thousands of years indicate that humanity has actually degenerated from cultural and universal prominence. Do you think the Empire State Building in New York will be standing in 100 years?

I think most of us would agree that the purpose of humanity to develop into more spiritual, loving beings. Our survival as a race of beings is totally dependent on this development. Not only must we learn to love each other, we must learn to respect the whole planet and everything on it and in it as our mother. It is interesting to note that many religions around the world teach the philosophy of going towards the light and/or becoming one with the light. The central message of Christian philosophy is becoming one with the son-sun-Christ. Christ is one of the human images of love and spirituality. The correlations here are no accident. Christ embodied the feminine aspect of being. His most basic message was very simple. No human being could experience the divine realms of consciousness without learning to love and honor the divine within him or herself and within every aspect of the world. We are moving towards the light or the sun in every aspect of our existence. We are moving towards the sun in terms of planetary evolution. We are moving towards the sun in terms of our understanding of harmonious life and energy resources (solar power is the most harmonious energy resource in the material universe).

Christ prophesied about a new Earth after the Tribulation...The Mayan prophecies point directly to Venus. The Christ image of love manifested in Central America as Quetzalcoatl. After 15 years of research, I am convinced that Christ and Quetzalcoatl are the same universal being. And now, I am finding other links to this same being in many other cultures as well. The Mayan legend of Quetzalcoatl is that, after many, many years of sharing the light of love with Central Americans, he was tricked by some jealous and envious people into an intoxicated state and was then tricked again into committing an immoral-sexual act. Quetzalcoatl became so ashamed of his deed and of his humanness, that he told the Mayan people that he had to leave this planet for a while. He had spent a long time with the Mayans, trying to help them create a beautiful civilization based on love and peace; but the people were not ready to enter the spiritual realms of Venus-love. Quetzalcoatl told them that he was going to Venus to prepare a new world! The last eclipse over the Mayan temple site (1991) corresponded with the prophecy that humanity will awaken to the experience of universal being and love. We are now being prepared for the next giant leap in our evolution. We are heading for Venus...

In my first book, I was beginning to understand the universal process going on here in this solar system. In "The Awakening of Red Feather" I wrote:

"In Christianity, Christ is synonymous with love. In all traditional mythology and astrology, love is synonymous with Venus. Someday the Earth will be dry and barren just like Mars for the Earth is surely changing in that direction now. Where will the human soul go then? Venus would seem to be the likely answer. Who can say that Venus is not ready for us now? If we were in spirit form and not in bodies, we might not feel heat and cold the same as we do now. In actuality, Venus may be ripe and ready for the collective human soul in spirit. The intense heat and low pressure atmosphere on the surface of Venus may be exactly what the soul of humanity needs to cleanse and purify itself...not unlike a giant collective sweat lodge."

Humanity is unconsciously longing for the warm and protective womb of creation. The Earth used to be warm and fertile and we have deep memories connected to that time. Could it be that "Quetzalcoatl"- "Christ" has prepared another birthing womb for the soul of humanity. What does it matter if we are in body or spirit? It is all the same...we just don't realize it in the here and now.

Venus is the brightest star in the morning and evening skies. It's there at those magical times of transition between the conscious and the unconscious dream-state to remind us of the love that our Creators have for us and that we should have for each other. Venus is there to instill peace and faith in our hearts. She represents the eternal Goddess and we are moving towards her. Humanity is moving towards a more feminine way of intuitive being. So many people are beginning to dream of a world of universal love. Our dreams are slowly becoming our reality. In times of transition, the old ways of being (in this case, the Martian ways of masculine domination and control, fear and insensitivity) resurface and manifest in violence and chaos. That is exactly what is happening on this planet today and is what has been happening on this planet for two thousand years.

So here we are...on the middle planet trying to find the balance between the male and the female. We are trying to find the balance between Mars and Venus. Are we going "back to Mars" or "forward to Venus"?

Chapter 6

The Return of the Goddess

I dedicate this chapter to Venus and Demeter, the Goddess of eternal life, love and nature...and I pray, with all my heart, that women regain their sacred ways of knowing and initiating human souls into the realms of universal love and respect for all life. For so many long centuries, the Goddesses of women and the feminine aspects of being have been denigrated and oppressed. For ages, masculine and patriarchal dominant cultures (symbolized by Mars) have repressed myth, mysticism, nature and the feminine ideas of oneness. To understand this oppression is to understand the purpose and destiny of human beings on Earth. It is to understand why the tree of Western scientific and intellectual knowledge precariously stands on entangled and suffocating roots. This oppression has not only robbed and pillaged Venus and the true power of women, but has enchained men as well in cycles of fear and destruction. And it has kept a stranglehold on the spiritual development of all humanity.

I have always tried to move my understanding of human existence forward in light and faith and I have always been entranced by the brilliant presence of Venus in the morning and evening twilight, not wanting to continue to explore traditional history and the sometimes dark and desperate attempts that men have made to explain the world - explanations that have denied the Mother Goddesses of the Earth. The ideas of the "self", the supreme male which has been defined as separate and individual may someday be seen as a great farce. Church philosophers and other rational warriors labeled the feminine ways of knowing as mythology that was evil and dangerous. It has been the attack and dismemberment of Goddess philosophy that led humanity into its current state of spiritual desperation. The legacy of Mars was and is still gripping the heart and soul of human destiny. Demeter and her followers would never have brought the world to the brink of social and environmental destruction. No, instead they would have created a planet sized garden of love, harmony and peace. They would have brought us into an age of universal being. The attempted destruction of Demeter was a terrible thing and the voice of Demeter is calling out loud and strong from the unconscious depths of humanity saying: "Give me back my daughters and give me back my dignity and honor so that I may help the sons and women slaves of Apollo avert a terrible fate of famine and mortal extinction."

In light of the desperate state of affairs on Earth, the myth of Demeter is a very important myth for all people to understand. The destruction of Demeter and the other Goddesses was fundamental to the formulation of rational, patriarchal philosophy that evolved into the political theologies of the Roman Catholic Church. Between two and four thousand years ago, many cultures around the world experienced a very similar shift towards male control and domination. A great fear of the feminine developed in the collective soul of humanity. The myth of Demeter tells this story very clearly:

In ancient Greek history, Demeter was a Goddess of equal or higher status with the male God Apollo. This was a time when women, in all respects, shared equal power and rights with men with the exception that women, not men were believed to be the ones holding the keys to eternal life. The Goddess had many names, depending on the culture; but, to the Greeks, she was Demeter. She was the Goddess of nature, crops, agriculture and cyclic rebirth for she embodied the mysteries of death and regeneration. She founded societies of women, and the spiritual energies of these societies produced fountains of eternal life. It was a sacred world of women's ceremonies that kept the powerful, rational male energies on the planet in balance and provided a direct link with the divine aspects of the universe. No Greek person, no matter what sex or walk of life, could find happiness and eternity without being initiated into her rites. The world of Demeter was a world of inner knowing - intuitive being. It was a world of natural beauty, harmony and sexual and cultural union. It was a world where individual separateness had no reality. All were one and everyone shared in the abundance of Mother Earth in equal measure.

Demeter had a virgin daughter named Persephone who was somewhat inclined to stay a virgin in the Goddess world, which was the ultimate expression of female empowerment and feminine experience. But she was abducted against her will by the male Gods and taken to the underworld to be the wife of Hades. The underworld was a deep, almost unconscious place where humanity had been trapped for ages...a place where the primal, physical and emotion aspects of material existence (wild passion, lust, greed, envy, jealousy) were untamed and glorified...a place were the male-Martian energies were extreme and unbalanced with the female energies. There was no receptive experience of love in that world; yet the masculine forces yearned for the love and freedom that Venus and the feminine realms brought to the planet. After the abduction of Persephone, Demeter lapsed into a deep period of mourning and withheld her love and blessings from the Earth. A great famine resulted which lasted for many years, bringing humanity to the brink of total devastation. Finally the male Gods returned Persephone to Demeter and the crops began to grow as a great season of rebirth (spring) fell upon all life. This set the stage on Earth for the Mars-Venus struggle that has characterized human experience for thousands of years. Persephone was raped and the male Gods had forced their way into the sacred realms of feminine being rather then honoring and patiently receiving the love that Demeter offered to all initiates. It's the same drama on the personal and intimate level of male and female relationships. The lessons for men have always been to open up to the inner, emotional and spiritual worlds of women and thus be initiated and invited into her being rather then to forcibly penetrate her sacredness. The world of the feminine is not just her body. It is also her mind and soul and her way of experiencing the world.

From this drama and struggle between the male and the female emerged the Eleusian mysteries which were the sacred and secret rites of women: rites of ceremony, fertility, and celebration for nature's cyclic rebirth from the depths of death and darkness. These rites were crucial to the developing spirituality of human beings and it was the responsibility of the Goddesses to teach all human beings to honor and keep sacred the garden of Earth. The Goddesses were openly active on this planet for hundreds and thousands of years in ancient times until the dawn of political Christianity and the shift in collective human consciousness towards feminine repression...a shift that can be seen in cultures all around the planet.

It is true that the ancients believed that men and women were different and that women were connected to nature and eternal life because of their ability to give birth and to continually regenerate their bodies every lunar cycle. The early Greeks worshipped Demeter seriously for they believed that they could not find peace in the afterlife without experiencing her world. The city of Eleusia was a real place in Greece and hundreds of thousands of people traveled there to seek the Goddesses. The ancients never questioned the concepts of male and female Gods and Goddesses. They believed that human beings were created as male and female in the image of our Divine Creators who must, obviously, be male and female! The ancients also saw the eternal and cyclic nature of life on this planet...the constant process of death and rebirth that women seemed to embody. They saw a process of universal love manifesting here in the material world. Societies could pillage and destroy the forests and fields but they always regenerated with beautiful flowers, berries, birds and animals. Women would bleed and a part of them would die every moon but they would always return healthy and fertile. The Goddesses were the foundation of all beliefs, and concepts of death were understood as processes of transition and rebirth.

The Demeter myth exemplifies how the power and divinity of women and the feminine has been, throughout the ages, stolen, repressed and controlled which has caused great harm and karmic retribution for all of humanity and the natural world. The central theme of this myth is very relevant to the current global crisis. Faced with total environmental destruction, the forces of patriarchal, masculine and political domination, still cling to their theories of supremacy over nature, animals and other human beings. They still think that it is better to force the natural, feminine world into submission rather than to work in harmony with it. Human beings have not only been destroying ever-larger tracts of land on this planet, they have been mismanaging the crops harvested both naturally and agriculturally. There has yet to emerge a fair and equitable system of food sharing and distribution on this planet. The Goddess Demeter, who I believe is alive and well, is not happy about the current state of affairs on Earth and she is, once again, beginning to withhold her love and blessings. I believe that she trying to warn us through the crop circles that are appearing at an increasing pace all around the planet. She is marking the agricultural fields in some mysterious way as she prepares for the coming years of famine and mortal destruction. The feminine, creative forces of the universe are trying to get humanity's attention by marking up our primary food sources. Many of the symbols and pictures that are manifesting in the world's grain fields can be seen as a direct link to ancient knowledge and Goddess worship! The first crop circles were simple, independent circles sometimes with interior designs.

Now they have evolved into complicated, multiple circles that are linked together with bars and triangles. It appears that some mystical-feminine force is trying to tell humanity to begin linking together and pay more attention to our grain harvests.

Generally, the women of the world are still in a state of repression and dishonor as the feminine aspect of being and Goddess worship is still considered an "evil" thing by dominant, intellectual and rational warriors. Historically, women have been denied the right to their Goddess ceremonies. They have been denied the right to explore and practice their own powers and truths, which resulted in an unbalanced Mars-Venus...male-female human situation on the planet. The oppression of women's truths took a terrible and desperate turn in Europe during the Dark Ages. It has been reliably estimated that 9 to 12 million women were burned and tortured for expressing the mystical and intuitive aspects of their natural, biological connection to the world. In China, Confucianism repressed the feminine to the extent of tying the feet of young girls so that they would never grow and mature, thus enslaving them to their husbands. In Central America, the Native cultures that were once very feminine and rooted in the Goddess, experienced the resurgence of male domination and repression that resulted in the Aztec priests literally ripping the hearts out of young women and throwing them down in cisterns. Modern societies have come from whole generations of women who were taught to fear their own inner, goddess nature. What we are seeing on this planet today is the attempt by the collective unconscious of humanity to heal this terrible imbalance. But to heal from centuries of tortuous abuse and oppression is no easy challenge.

One of the most effective forms of repressing women is by controlling and manipulating their sexuality. Women are very much connected to nature and the cosmic regenerative forces through their sexuality. No human soul can manifest into this material world other than through the power of women. In ancient cultures, this power was considered the most sacred magic of existence. But, in many cultures, the men developed resentment and jealousy for this power and thus tried to claim it as their own by attempting to control, manipulate and dominate women. Extreme examples of this negative behavior are everywhere, ranging from the millions of cases of female genital mutilation in Africa and elsewhere every year to the large scale absence of midwifery in the "modern" world. Childbirth in "civilized" societies has been stripped of its sacredness and put under the cold scalpel of masculine-scientific observation and control, even to the point of creating myths about the mother's milk being harmful to the baby! Such myths were created for controlling women even further as many people have realized in recent decades. What better way to control women than to take away from them the very biological connection and power that is central to their natural experience of the world, thus leaving them in a miserable state of spiritual and material dependence. As stated before, the procreative power of women was considered sacred and magic and women held social positions of honor and leadership in almost ancient cultures. Recently, women have made great strides in gaining positions of leadership in the modern, male-dominated world; but, in many cases, these positions are defined by the norm of intellectual and masculine control. The Venus concepts of love, harmony, material equality, cultural union, peace, environmental respect, trust and forgiveness have not had a chance to manifest in great measure anywhere on this planet. There are changes in the wind though, for the

Goddess is resurrecting from the depths of the human psyche.

Yes, the Goddess is returning and the question is: will the men that are in power in this world embrace her return or try to trample her back into the shadows and back into the unknown and unrealized mystic realms? To be woman or feminine...to be intuitive, mystical, sexual, natural, and ceremonial is not "bad" or evil. All human beings, especially men, have a deep yearning for these feminine qualities. There has been so much repression of the feminine in recent historical times that very few people in the world have been educated by or initiated into concepts of feminine being. And most all of the modern world is somewhat enchained by the male-Martian forces that are built right into our social structures, our language, our value systems and our religious, political and educational institutions. Human beings have just begun to question these systems and institutions. As the next generation unfolds on this planet, we will likely see the abandonment of most of these old, controlling and fear-based ways of being.

Late in the fall of 1994, a white buffalo was born in the heartland of America. For many Native American people, this event was equivalent to the return of Christ in Christianity. The birth of the white buffalo is the return of the Goddess in one of her many forms. The story told by Sioux medicine people reveals the need of men to honor and respect the feminine aspects of being in this world...

The Goddess came to the Sioux as White Buffalo Calf Woman. She appeared to these Native Americans at a time when more and more conflict and war was developing between the plains tribes and shortly before the onslaught of the Europeans. It was said that she brought a sacred pipe of peace and that human beings would have to honor this pipe before time could be measured - before the end of this age. She came at first as a white bison according to the story, but she appeared to two warriors as a beautiful woman coming along in a meadow. One of the warriors had lustful and selfish thoughts, but the other one was full of awe and respect. She told the first one to come and do what he wished with her and he did. When he was done, he became engulfed in a fog. Soon the fog lifted to reveal nothing but a skeleton that had snakes withering in its bones. She told the other warrior to go back and prepare a place for her in his village for she was coming with the pipe.

There are a few different versions of this
story but they all speak of White Buffalo
woman as a magical and love-centered being
that taught the principles of the spirit within:
compassion, peace and harmony - much the
same message as Christ brought.

It's a powerful story that pretty much sums
up the condition of human experience on this
planet. Just as the first warrior was reduced
to dust and bones, so, too, will modern humans be reduced to dust and bones from their
own self-created famine, wars and environ-
ment destruction if they continue to repress
and dishonor women and the feminine. Our
future and our hope for survival is in everything that Venus and the Goddesses represent, even if we never experience the physical
reality of the next planet towards the sun. I
am inclined to believe that human beings will
walk on the surface of Venus sometime in the
future, but we may need to evolve into more
spiritual beings and lighten our material
bodies to some extent first. The practice of not
eating animals is one of the most important
transitions that most of humanity has yet to
make. The eating of animal flesh has kept
human beings firmly grounded in the material
world and caught in the cycles of physical
being. The times of natural wildlife abun-

dance are over on this planet, at least for the present. The times of primal, aggressive male behavior, which has included the domination and control of animals, nature and women, must also end.

All human beings, women in particular, must be given the freedom to create new concepts, norms and social standards that express the original ideals of feminism and the Goddess. It is an absolute moral responsibility of men to give women the opportunity to re-connect with the mystical and intuitive realms of creation. Men must begin to connect with these realms themselves. We must revive the myths of feminine-Goddess experience. Myths are powerful oracles of society and they reveal the plight of humans in this dimension. The Christian "myth" of Eve in the creation story is a perfect example of how social and political power can be given to one sex over another through myth. The ancient Hebrews and later the Christian theologians were very clever and manipulative with language and symbols. The symbol of the snake, which in ancient cultures represented a very divine, feminine power of death and regeneration, was shown to be evil and detrimental to people. The symbol of the snake was always connected to healing and female energy previous to early Christianity. We were taught to fear the things that the feminine symbols represented. We were taught to fear the mysteries of life...the fruits from the tree of feminine knowledge. So much time and energy has been wasted by humanity in trying to repress the feminine energies. We come from the feminine and we exist only through her body whether it be our biological mother or the Mother planet.

Yes, I pray for the wisdom and power of Demeter and the other aspects of the Goddess to fill the hearts and minds of all those in search of happiness and universal truth. The virgin world of Venus awaits all those who honor the Goddess and accept the necessary transformations of love, forgiveness, humility, brother and sisterhood. We must never forget that all human beings are children of the Blue Earth Star. Whether we like it or not, we are all in this together. Let us work together to prepare a place within our hearts for the return of the Goddess. Let us make this a sacred altar where no fearful, violent and aggressive male energy can enter. Humanity is braced to enter a new age. Our leaders, prophets, healers and teachers will be those who have found harmony with both the male and female within. It seems that nothing is sacred in this world any more. These are the lessons of Earth and human beings will never evolve beyond the current state of self-destruction, no matter how much technology they have, until they find the balance between Venus and Mars here on Earth.

Chapter 7

We are Beings of the Universe!

We are universal beings that have been sent or chosen to come to this solar system to exist for a time. We are the expression of whatever our Divine Creator force is. We are so much more than mind and body. We are so much more than soul. But we are nothing at all if we never have the opportunity to know, feel, touch and experience our beingness in all its universal forms. The physical and material realm is only one of the many realms of existence. It is only one of the spectrums, wavelengths and vibrations of light and energy. I had a vision once that there were seven basic spectrums of light, seven basic wavelengths of energy and seven basic vibrations of sound. What we experience as the material world is a blend of one spectrum, one wavelength and one vibration. But there are as many realms as there are possibilities of mixing the different combinations of the three sets of seven. This was my vision, but that does not mean that I take this revelation as literal. There could be many other ways of seeing the metaphysical nature of existence. But I learned from this vision that human beings have not even begun to experience all the realms of being that exist in and around us.

The solar system is connected from the core of the sun to the outer orbits of Pluto and beyond. Gravitationally and electro-magnetically, it is one body. The elements of Jupiter or Saturn are the same elements that exist on this planet and in our bodies. Human beings are able to exist on Earth only by the continued gravitational stability of all the planets. If Mars, Jupiter or any one of the planets were to explode or somehow enter a new orbit around the sun, every other planet in the solar system would have to re-arrange to a new gravitational balance which would change the material and conscious conditions on Earth dramatically. The solar system is existing in a interdependent field of magnetic, gravitational, electrical and anti-material forces. Here on this planet, human beings also exist in all these realms. We are more than physical body. We are more than electric-magnetic minds. We are soul...spirit - what scientists are now calling anti-matter. Our physical, mental and spiritual bodies are connected to every aspect of motion, balance and energy wave of this solar system. If there are large eruptions on the surface of the sun, life on Earth will experience electrical and magnetic disruption. Jupiter emits powerful radio waves. If there are explosions on Jupiter's surface, (as there were a couple of years ago) then there would be changes in the energy frequencies on Earth and on all the planets. In traditional astrology, Jupiter represents mass religions, organizations, and governments. The positive side of Jupiter is spiritual development, expansion of the soul and the expansion of human knowledge. The negative side is religious, fundamental fanaticism, bigotry, close-mindedness and obsession. Look at the world around us. Are we not experiencing major changes in all these aspects of human understanding? The explosions on Jupiter caused by the impact of asteroids have stirred up the spiritual condition of humanity and have set in motion disrupting waves in our conceptual base of reality.

Astrology is really the science of understanding the electrical-magnetic, gravitational and elemental relationships between Earth, the planets which affects human conscious and unconscious behavior. For example, Saturn has always symbolized restraint, form, law and control. Methane, ammonia and hydrogen are the predominant elements of Saturn and Saturn's existence in this solar system represents the a certain control and balance of these elements. Methane is a very potent and toxic gas usually encountered by miners deep within the Earth. It is a very important substance in the chain of life but must be controlled and restrained. Thus, the astrological nature of Saturn emerges from its physical and elemental reality of being. Saturn also emits subtle radio waves that connect all the atoms of methane and ammonia in this solar system. If Saturn's orbit comes into gravitational stress, one could predict that methane and ammonia levels in or on this planet would become more dangerous and explosive. And this same elemental relationship extends into our bodies and our minds. Human consciousness would also be affected by Saturn's gravitational stress.

Not only are human beings directly connected to every aspect of this solar system, we are also directly connected to the galaxy and beyond. Our solar system is a part of a larger body of being, and changes in that body will create changes in ours. If our solar system moves into an area of the galaxy that has higher or lower vibrational frequencies, our bodies and minds will go through intense periods of change and our experience of the material-physical world will shift. This is why the Mayans considered it so important to study eclipses. The eclipse cycles reflect universal changes. They were one of the only cultures to actually record the movement of our solar system into different areas of our galaxy...and the movement of our galaxy into different areas of the universe. The Mayans actually understood that we are universal beings and they predicted the end of material time and the material world as we know it now! I have come to believe that all human experience on the surface of this planet is reflecting everything that is happening in the universe. If there is conflict here on Earth, you can bet there is conflict somewhere out there in space. If we are truly created in the image of our Creator and we are children, parents, grandparents and great-grandparents, then you can bet that the same relationship exists between the beings of universal creation- our Gods and Goddesses.

Who are we and were did we come from? What is the true nature of our Creators? There is no way that these questions can be answered through our current levels of con- scious experience as human beings. We have been trapped in ignorance about the nature of our Creators and the universal process of existence ever since we began to view our- selves as separate from nature and the cycles of being that continually evolve around us. When our ancestors began to repress feminine concepts and universal ideals, humanity lapsed into a dark age of isolation from universal being. But we are awakening now in great measure and at an accelerated rate. Soon, there will likely be many mystical and spiritual happenings on the surface of this planet that no one could have ever imagined. There already have been. And, in the next five years, there will be changes in consciousness so dramatic that human beings may discover a whole multitude of realities and worlds all around us. We may discover from where and from what we came. As our planet changes and as the social, political and economic condition of humanity continues to worsen, human beings will be forced to release patriarchal conceptions of reality. They will be forced to look towards the skies, to other planets in this solar system and into the greater realms of space. Human beings can only see what they have been conditioned to see. They can only sense what they have been conditioned to sense. For two or three thousand years, most of humanity has been conditioned to fear the intuitive and mystical knowledge of the world. We have been conditioned to fear our own Creator and divine past. Thus we have created a world that reflects our spiritual condition of repression, control and domination.

Our Creator is not some angry, judgmental male god sitting on a throne somewhere in the heavens. What a ridiculous conception! We are the reflection and manifestation of a process of creation that is both male and female...our Creators have creators and the process comes all the way back to and through us. Long ago human beings were entrusted with the garden of Earth to experience creation on more intimate and emotional, physical levels. Our experience here will determine another level of our re-creation somewhere else. There are so many worlds out there, billions of planets where some form of existence is going through the same process. But we Earthlings went down a path of separateness, male domination and fear of universal connectedness and have been unable to get beyond our own noses and our notions of the Earth as the center of the universe.

There are many conceptions of creation, Gods and Goddesses, and there are many conceptions of the afterlife. Who can say what our Creators are really like and what happens to the human soul during the transition that we so often call "death"? Most people are in agreement that our souls continue to exist on some level; but the truth is we have had very little understanding of the nature of existence and the realms beyond the ones we presently sense. The collective human consciousness is beginning to change and the stage is set on this planet for the greatest transformation of 'being' that humanity has ever known. The more we open our minds and hearts to the universe and to other realms of existence, the more we will be able to sense and experience these other realms right here in the present. This is exactly what is happening! Humanity is beginning to experience all kinds of non-material realities...UFOs, crop circles, alien creatures and other so- called "mystical" happenings. And the more our conscious-material world of masculine domination disintegrates, the more of the universe we will be able to perceive. There are many people in positions of political and religious power that will surely condemn the message of this book as they try to consolidate and hold on to their selfish, destructive worlds. But there are very advanced spiritual and universal forces surrounding this planet and these forces are appearing through the hearts of open minded people and they will not allow a return to the dark, repressive ages of the past. Humanity is heading for a new age of love, peace and universal being.

We must come to realize that this planet is a living being that pulses with the creative forces of the universe. Everything is connected. This is no separateness anywhere on Earth or in the cosmos other than in the concepts and ideas of human beings. But these concepts and ideas have been given a lot of power and have created false, destructive realities. There is no separateness between humans and other humans, humans and nature, humans and the universe or humans and their Creators. In each of our bodies there are billions of atoms, millions of untold elements that have flowed around the universe and around the surface of this planet in constant flux since the beginning. The same atoms or elements that were once part of a giant redwood tree five hundred years ago or once were a part of an Australian Aborigine's body 1,000 years ago are now a part of my body; the movement of water has linked all life on the surface of Earth. And these same atoms and elements were once another part of this galaxy or the universe.

Human beings cannot continue to live unbalanced with the Mars-Venus, male-female aspects of being. We cannot continue to act independently and selfishly, while acquiring the material wealth of this planet for ourselves when others live in material desperation. We must strive to create a more balanced, tolerant and peaceful world in all respects. If my brothers and sisters are suffering, then so am I. If the Earth is wounded and losing its ability to renew itself, then so am I. If the plants and animals disappear from the planet then so will human beings. So here we are, on Earth- the middle planet between Mars and Venus...

Where do we go from here?

ABOUT THE AUTHOR

Jonathon Ray Spinney is an artist, sculptor and metaphysical philosopher who spends time in both Maine and New Mexico. His philosophy is very simple and Earth based: we are here to learn the lessons of love, honor and dignity for all life during our journey on this blue star. The spiritual condition of our soul is all that matters in the end...

Other Books by Jonathon Ray Spinney-

"The Awakening of Red Feather- Dream, Prophecy and Earth Changes"